The Magic Of Science

Dennis Regling

Magic Marketing Co; The Pickle Group, LLC
Freeport, Ohio 43973

To my daughters, Alicia, Heidi and Mara

"Success is going from failure to failure
without a loss of enthusiasm."
Winston Churchill

"In order to succeed, your desire for success
should be greater than your fear of failure."
Bill Cosby

The Author
"Mr. Dennis"
Dennis Regling

TABLE OF CONTENTS

INTRODUCTION

You may be wondering what magic and science have in common.

The answer is everything.

Every time you have ever seen a magician, on a stage or on television, every trick uses science.

Long, long ago, the very first scientists were the magicians, and the witch doctors. When they knew something that others didn't, they could make it look like they had great powers.

In this book, we are going to look at some of the ways magicians use science to fool us. We will do some fun experiments and you will learn some fun ways to fool your friends.

"Knowledge is power."
Francis Bacon

CHEMISTRY

So you're asking, what is CHEMISTRY?

Chemistry is the study of MATTER and the changes that take place with that matter.

Don't ask why that matters. It just does. A lot. Everything around you is made up of matter.

Your body, your furniture, even the air you breathe are made of matter. Everything on Earth, everything in our solar system, everything in our galaxy, and everything in the universe is made up of matter.

Matter is the name that scientists have given to everything that you can touch, or see, or feel, or smell. Another way to put it is, matter is anything that has mass and takes up space.

Here are a couple of experiments that can make chemistry look like magic.

WARNING:

Before doing any experiments, get your parents' permission. You may also ask them to help.

When doing chemistry experiments, always use the proper safety equipment.
- Use safety glasses to protect your eyes.
- Wear a smock or a lab coat to protect your clothing or wear old clothes.
- Use gloves when handling chemicals with your hands.

CHEMICAL REACTION MAGIC

What You Need:
* 3 empty plastic bottles or clear plastic cups - **no lids**
* water
* dishwashing detergent
* baking soda
* white vinegar
* Cherry Juicy Juice brand juice

Get 3 empty bottles like Gatorade drink bottles or other wide mouth bottles. Use plastic, not glass. You can also use clear plastic cups.

In the first bottle, put 1 TBSP of baking soda and ¼ cup of water.

In the second bottle, put ½ cup of Cherry Juicy Juice. Cherry juicy Juice 100% juice works best for this experiment. You can also use grape juice.

In the third bottle, put ½ cup of white vinegar.

What's the difference between chemistry and cooking? In chemistry, you should never lick the spoon.

A Little Chemistry Magic

Color Changing Liquids.
We know when you add red and white, you get pink.

Take the first bottle, with the baking soda and water. Shake it to mix the two. You will have a milky white liquid.

Now, add the juice to the white liquid. We would expect it to turn pink. Instead, it will turn green.

Looks like Frog Juice.

WHOOSH !!!

Now add the vinegar to the Frog Juice.
It will go WHOOSH. Like a volcano, it will erupt over the top of the bottle.

Always be sure to do this over the sink, over a bucket or out of doors.

For more fun, have a friend add the vinegar to the Frog Juice - they will make a funny face.

For a BIGGER WHOOSH,
Add 1 TBSP of dishwashing
detergent to the baking soda and
Water BEFORE adding
the vinegar.

Explanation

This experiment can be made to look very magical with the right presentation.

You know that it is just chemistry. The color change and the whoosh are chemical reactions. A chemical reaction is when two or more chemicals come together and become something else.

The baking soda and water form what scientists call a **base**.

Fruit juice, like cherry juice, grape juice, apple juice and orange juice contain an **acid** called citric acid.

Did you know you drink acid for breakfast? Some acids, like fruit juices are good for you. Never drink battery acid or bath water.

Vinegar is also an acid.

When an acid combines
With a base, there is
always a chemical reaction.

Never mix chemicals if you do not know what the reaction will be. Some chemical reactions can be dangerous. Always read the label before mixing cleaning supplies like soaps and detergents.

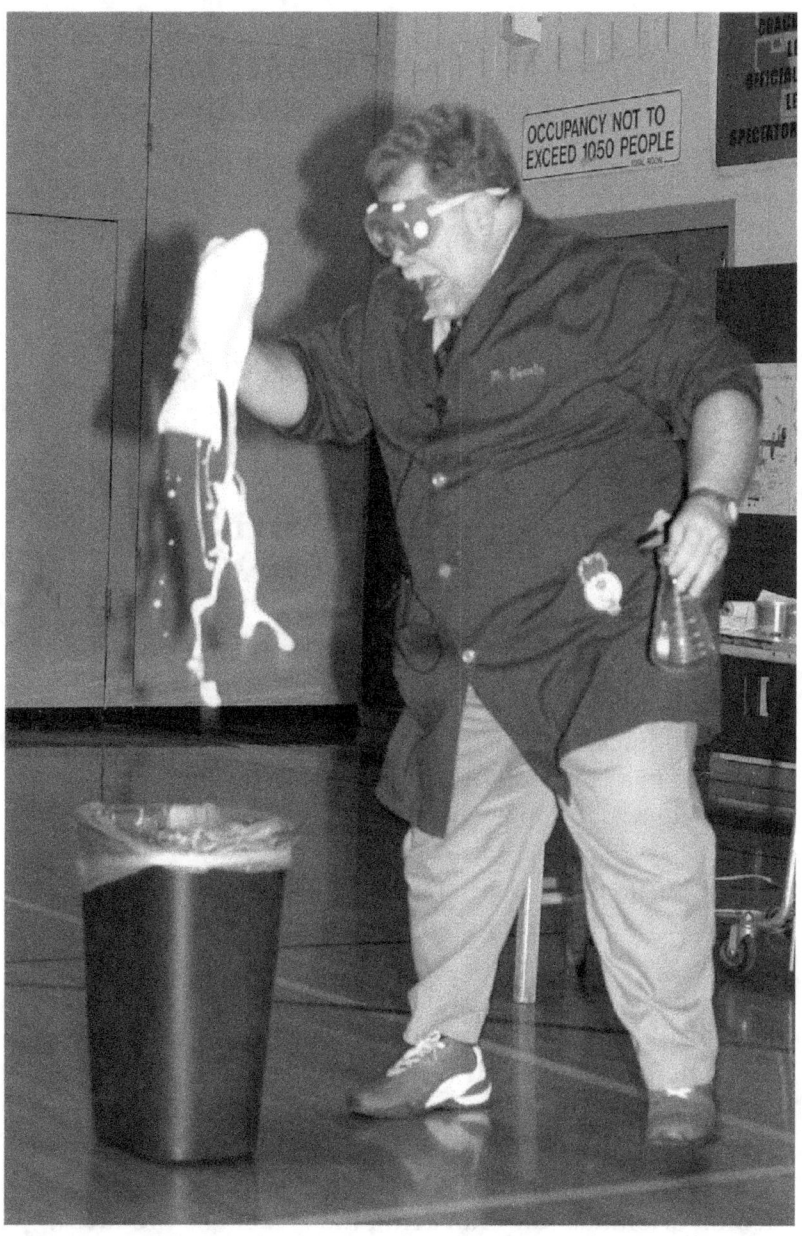

Author Mr. Dennis discovers that baking soda and vinegar result in a messy chemical reaction when mixed. Mr. Dennis is always careful to wear his safety glasses and a lab coat .

Mr Dennis' Almost World Famous Homemade Orange Soda-Pop

This experiment's a real gas! And you'll see what I mean when you make a bubbly orange soda-pop that's delicious and good for you too.

What you'll need:
  * half glass real orange juice (or lemon juice)
  * half glass water
  * 1 teaspoon of baking soda
  * sugar or some other sweetener

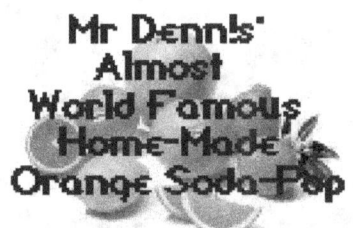

What to do:
  1. Add juice to water.
  2. Stir in a teaspoon of baking soda. Now take a sip; add some sugar if it's not sweet enough.

What happens: Your homemade soda pop will be fizzy and taste like lemon or orange soda pop.

Why this happens: All that bubbling is the result of a chemical reaction between a base and an acid. A base is a compound that gives off negatively charged hydrogen ions when dissolved in water, and an acid is a compound that gives off positively charged hydrogen ions when dissolved in water. When you mix the baking soda (a base) with the lemon or orange juice (an acid), they react and produce a fizzy gas called carbon dioxide. (The bubbles in store-bought sodas are also created by carbon dioxide, but they're added under pressure to water.)

DIAPER CHEMISTRY

Ask the mother of a baby for a disposable diaper.
Make sure it is a clean diaper.

Cut open the diaper and hold it over a newspaper. A powder
will fall out of the diaper. This powder is called Sodium
Polyacrylate.

It is perfectly safe, but be sure not to get it in your eyes.
Use safety glasses.

Sodium Polyacrylate powder is a super absorbent. I call it a
chemical sponge.

When water is added to the white crystalline powder, it absorbs
the water like a sponge. As it absorbs the water, the tiny pieces
get bigger and form a gel.

Disposable diapers use small amounts of Sodium Polyacrylate
to absorb baby urine. The more polymer powder in a diaper,
the more urine it can absorb.

In addition to its use in disposable diapers
Sodium Polyacrylate has many uses.

1   It is added to potting soil to help soil retain water.
2   It is used by florists as a dirt-free way to store water
        and to keep cut flowers fresh for a long time.
3   It is used in filtration units that remove water from
        airplane and automobile fuels so that the vehicles
        perform better.
4   It is used to make Gro-Creatures™, which are toys
        shaped like dinosaurs, fish, lizards, and other assorted
        animals that increase in size when placed in water.
        These can be dried out and used over and over again.

15

DIAPER GEL MAGIC

Get a styrofoam coffee cup and a handkerchief.

Put 1 TBSP of Sodium Polyacrylate from a clean diaper in the bottom the cup.

Do not let your friends know you have done this.
Now comes the magic.

Pour water into the cup with the Sodium Polyacrylate.

Cover the cup with the handkerchief and say some magic words. I suggest "Purple Puppy Chow."

Remove the handkerchief and turn the cup upside down.
It's Magic - the water has vanished.

The Secret:
When the water mixes with the powder, it forms a gel.
When you turn over the cup, the gel stays in the cup giving the appearance it has disappeared.

OPTICAL ILLUSION

Optical Illusions -the use of shapes, color, and line distortions which trick the eye and brain.

The term Optical Illusion combines two big words. **Optical** means your eye and **Illusion** means something that tricks you or fools you.

An **Optical Illusion** uses your eye to fool your mind. In other words what you do see you don't see and what you don't see you do see. You see?

Magicians are experts in creating optical illusions. Using space, light and movement, magicians fool your eyes into seeing what the magician wants you to see.

When you see a magician make a lady float, that is an optical illusion.

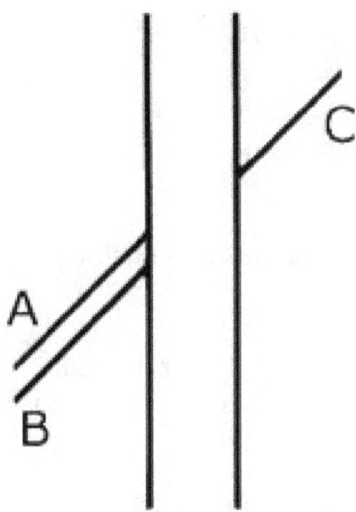

The Poggendorff Illusion is an optical illusion that involves the brain's perception of the interaction between diagonal lines and horizontal and vertical edges.

In the picture above, a straight A and B line is obscured by a "track". The A line appears, instead of the B line, to be the same as the C one. A ruler cab be used to show this is not the case.

Look at the picture to the right.

Do you see an old lady or a young lady?

You should be able to see both.

Do you see the seal? Do you also see the bear?

How many black dots do you see? There are none...

Mr. Dennis demonstrates an optical illusion.
Do you see a piece of pie
or do you see a pie with a piece missing?

MATHEMATICS

Did you know, you cannot have fun without using mathematics?
You can't play games and sports and keep score without using math.

You can't go to the store and spend your money and count your change without math.

You can't bake a cake or build a birdhouse without using math.

I looked it up. You cannot have fun without math. So if you want to have more fun, ask your teacher for more math homework.

Magicians also use math in their tricks. They use math to build magic boxes. They also use math to perform tricks that appear to be mind reading.

Here is an easy trick using math:

1   **Ask a friend to choose a number between 1 and 20.**
2   **Ask them to double the number in their head.**
3   **Add 6 and divide by 2.**
4   **Subtract the original number from the new number.**
5   **Pause for suspense and announce that the number is 3! Try it with any number. The answer will always be 3.**

MIND READING MATH

You will have your friend think of one of the animals on the next page. You are going to magically reveal the animal they are thinking of.

Instruct your friend to spell the name of their animal in their mind, (not out loud), as you point to various cards. When they get to the last letter in their word, they are to say "STOP."

It will appear you are pointing at random spots, but you will point as follows:
- First move - point anywhere
- Second move - point anywhere
- Third move - COW
- Fourth move - DOVE
- Fifth move - SKUNK
- Six move - BEAVER
- Seventh move - DOLPHIN
- Eight move - ANTELOPE
- Ninth move - BUTTERFLY

When your friend says "STOP," you will always be on their selection.

You should notice that each of your moves takes you to an animal with one more letter in its name than the one before. COW has Three letters. DOVE has four letters. Etc.

You should also notice that after you get to COW, you will be moving clockwise and skipping one square each move. If you remember this pattern, you do not have to remember the order of the animals.

SKUNK

ANTELOPE

BUTTERFLY

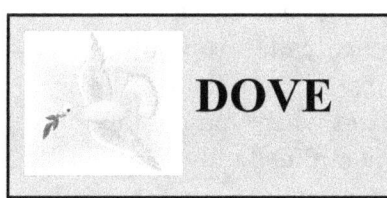

DOVE

BEAVER

DOLPHIN

COW

Mobius Magic With Shapes

This is an interesting mathematical trick

1. Cut a strip of newspaper 2" wide and 24" or longer.
2. Holding the strip out straight, give it a half twist (180 degrees) and attach the two ends together.
3. Take a pen and draw a line along the center of the strip.

You will discover you have a one-sided loop. Now cut the strip along the line you drew. How many loops do you get?

Your loop is called a Mobius loop, which is a shape described by a mathematical science called **topology**. When you twisted your strip, the inside and outside became one continuous surface. And when you cut the strip, it became one longer loop but still had only one continuous surface.

Now, try the experiment again, but this time give the paper a full twist. You'll be surprised by the results.

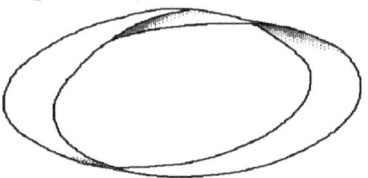

History: The Mobius strip is named after August F. Mobius, the German mathematician who discovered it. During the early 1800's, Mobius helped develop a study in geometry that became known as topology.

Topology explores the properties of a geometrical figure that do not change when the figure is bent or stretched.

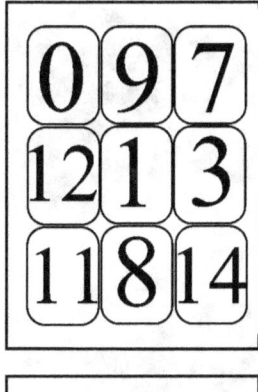

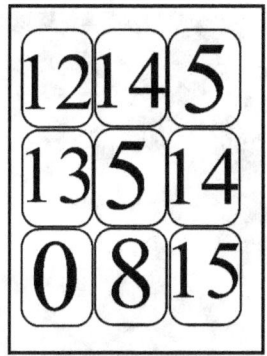

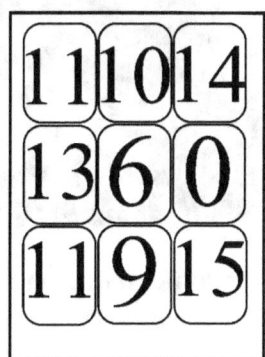

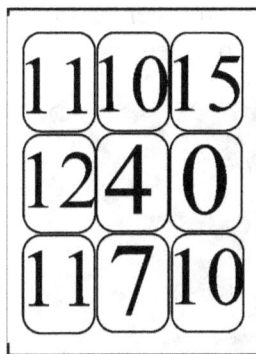

CALCULATOR MAGIC CARDS

Have a friend think of a number between 1 and 15. Have your friend tell you which cards their number is on.

By adding the middle numbers of the chosen cards in your mind, you will be able to tell your friend their number.

EXAMPLE: If the chosen number is 7, the number seven is on the three cards in the left column. Add the middle numbers of these cards. **1 + 2 + 4 = 7**

Mr. Dennis performs some mathematical magic with his set of calculator cards.

Mr. Dennis has performed his science show in over 1200 schools in 22 states. Students and teachers both enjoy his magic while they learn about science.

PHYSICS

Physics is the science of how things move and how things work. Physics includes such things as machines, magnets, forces, friction, air pressure, water pressure, wheels, levers, gears, pulleys and more.

Gym class is called "physical education," because we use physics in gym class. We use physics to play basketball, jump rope and to fall down.

Physics also has a big role in the history of magic.

Most people have heard of Harry Houdini. Harry Houdini was a great magician and an escape artist. He used his knowledge of physics to develop grand illusions where he would escape from chains, handcuffs, locked trunks and even safes.

What most people do not know is that Harry Houdini's real name is Erich Weiss. Houdini took his stage name from Jean-Eugène Robert-Houdin.

Jean-Eugène Robert-Houdin is considered the "Father of Modern Magic." He was a brilliant inventor and showman.

Jean-Eugène Robert-Houdin was one of the first people to find a use for electromagnetism. He created a new trick called "The Light and Heavy Chest." He would invite a small girl on stage to lift the chest. She would lift it with ease.

Then Houdin would invite a large man to do the same. No matter how hard the volunteer tried to lift the chest, he couldn't move it.

Hidden inside the wooden chest was a metal plate, and an electromagnet sat under the stage. When his assistant turned on the magnet, the strong attraction made it impossible to move the chest.

Robert-Houdin wrote in his autobiography that at this time *"the phenomena of electromagnetism were wholly unknown to the general public. I took very good care not to enlighten my audience as to this marvel of science."*

Mr. Dennis uses a Tesla Coil to pass electricity through a teacher and two students to light a fluorescent bulb.
Mr. Dennis has had special training to learn how to perform experiments like this safely.

Make an Electromagnet!

You will need:
* A large iron nail (about 3 inches)
* About 3 feet of THIN COATED copper wire
* A fresh D size battery
* Some paper clips or other small magnetic objects

1   Leave about 8 inches of wire loose at one end and wrap most of the rest of the wire around the nail. Try not to overlap the wires.

2   Cut the wire (if needed) so that there is about another 8 inches loose at the other end too.

3   Now remove about an inch of the plastic coating from both ends of the wire and attach the one wire to one end of a battery and the other wire to the other end of the battery. See picture below. (It is best to tape the wires to the battery - be careful though, the wire could get very hot!)

4   Now you have an ELECTROMAGNET! Put the point of the nail near a few paper clips and it should pick them up!

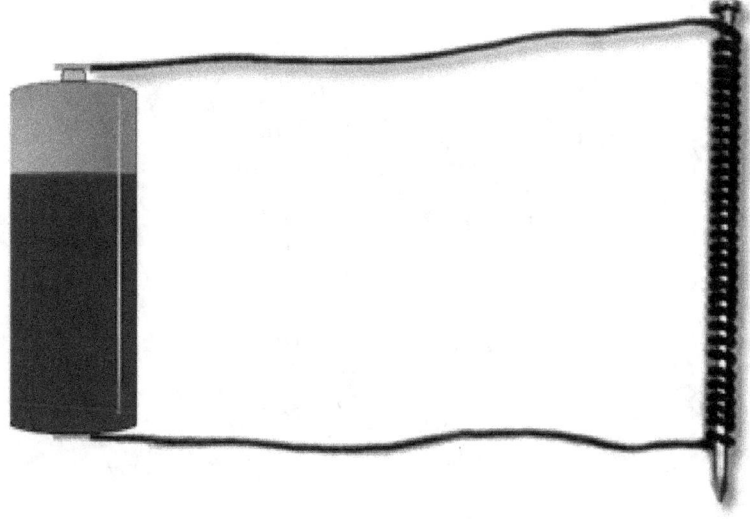

Most magnets, like the ones on many refrigerators, cannot be turned off, they are called permanent magnets. Magnets like the one you made can be turned on and off and are called ELECTROMAGNETS. They run on electricity and are only magnetic when the electricity is flowing.

The electricity flowing through the wire arranges the molecules in the nail so that they are attracted to certain metals. NEVER get the wires of the electromagnet near a household outlet!

Be sure to disconnect the battery when you are done experimenting. Be safe - have fun!

Nikola Tesla was an amazing inventor. His work resulted in over 1000 patents.

Nikola Tesla worked alongside Thomas Edison several years. Edison invented the light bulb and Tesla showed him how to plug it in.

Seriously, Edison did his experiments with DC or direct current electricity.

Tesla developed a system with AC, alternating current, that we use to wire our towns and homes to use electricity.

Tesla also discovered that electricity can be sent through the air without wires. His work led to the invention of the radio and more.

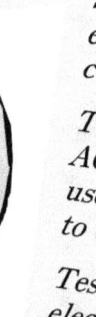

Make A Paperclip Float!

You will need:
- clean dry paper clips
- tissue paper
- a bowl of water
- pencil with eraser

1   Try to make the paper clip float. It won't float.
2   Tear a piece of tissue paper about 2" by 2"
3   Gently place the tissue flat onto the surface of the water
4   Carefully place a dry paper clip flat onto the tissue
5   Use the eraser end of the pencil to carefully poke the tissue (not the paper clip) until the tissue sinks.
6   Done carefully, the tissue will sink and leave the paper clip floating!

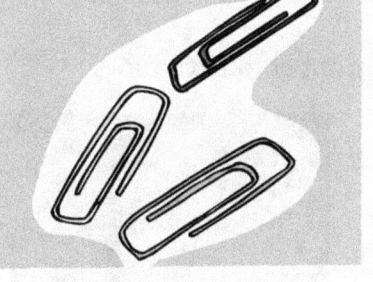

How does this work? The magic is caused by **surface tension**. Basically it means that there is a sort of skin on the surface of water where the water molecules hold on tight together. If the conditions are right, they can hold tight enough to support your paper clip.

The paperclip is not truly floating, it is being held up by the surface tension. Many insects, such as water striders, use this "skin" to walk across the surface of a stream.

CONCLUSION

By now, you should have read the entire book and performed many of the experiments.

You now have a basic understanding of how magicians use science to fool us. Next time you see something magical, or if you don't know how something works, go to your local library.

When you figure out the science which makes the magic work, you can no longer be fooled.

Your local library will have books with magic tricks, science experiments, math puzzles, and more. You may also want to look up biographies of Harry Houdini, Nikola Tesla and other great scientists and magicians.

Remember to have fun.

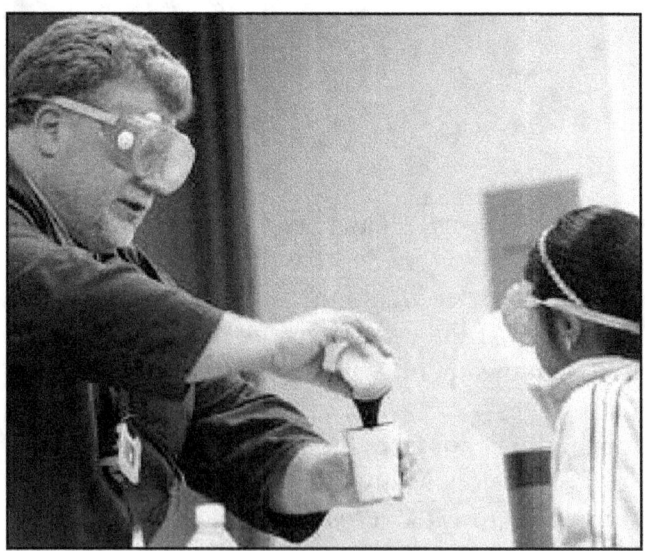

Mr. Dennis performing an experiment with a student.

RESOURCES

For more magic fun and safe experiments, go to:
http://www.magicscientist.com

To schedule an educational assembly for your school or group, contact Mr. Dennis at:
http://www.greatassemblies.com

For great teacher resources, go to:
http://www.magicscientist.com

IMPORTANT NOTICE
Mr. Dennis does not answer emails from students or children under 18 years of age.

If you have a question about one of the experiments in this book, have your teacher or parent email me at:
MrDennis@greatassemblies.com

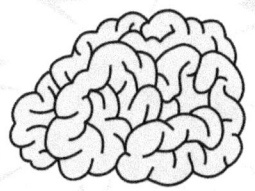

Remember to exercise your brain by reading every day.
"Knowledge is power."
Francis Bacon

ABOUT THE AUTHOR

"Mr. Dennis" Regling lives in Freeport, Ohio with his wife Karen and several cats. Freeport is a small town nestled in the foothills of the Appalachian Mountains.

Mr. Dennis has two magic rabbits, David Hopperfield and the Great Hare-dini. He has a dove, Clyde, who often joins him on stage.

Mr. Dennis has performed educational programs in over 1200 schools in 22 states. He also does youth ministry and stage magic shows.

In addition to performing magic, Mr. Dennis is a skilled balloon artist and is often seen performing at fairs and festivals throughout the Midwest.

Mr. Dennis has written 12 magic books and produced several instructional DVDs for balloon artists.

Mr. Dennis performs regularly at the Victoria Vaudeville Theater in Wheeling, WV and has also been on radio and television.

In his spare time, Mr. Dennis enjoys cooking, gardening and fishing with his wife, Karen.

NOTES

www.ingramcontent.com/pod-product-compliance
Lightning Source LLC
Chambersburg PA
CBHW072046190526
45165CB00018B/1844